Life Without Animals

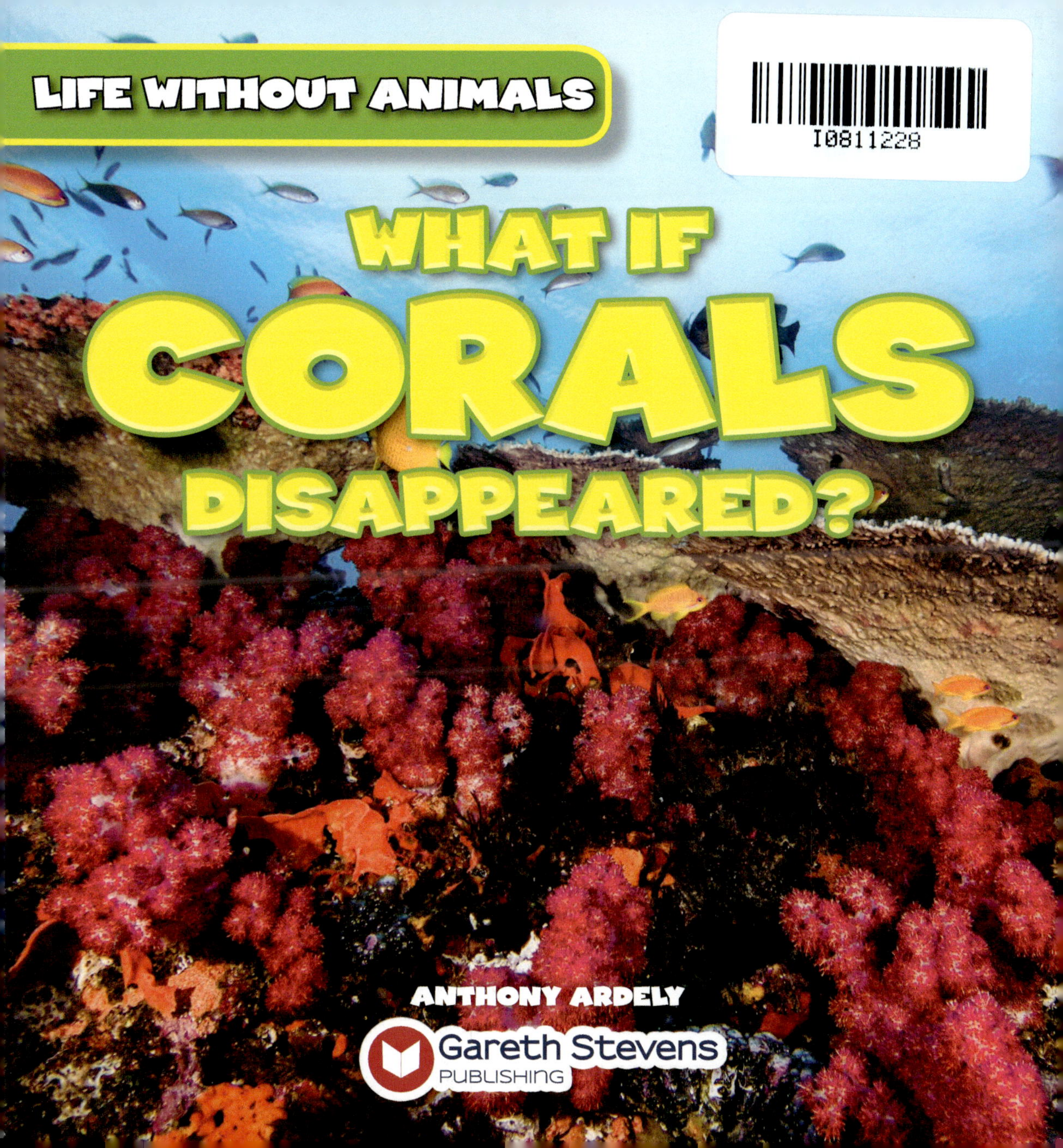

What If Corals Disappeared?

Anthony Ardely

Gareth Stevens Publishing

Please visit our website, www.garethstevens.com. For a free color catalog of all our high-quality books, call toll free 1-800-542-2595 or fax 1-877-542-2596.

Library of Congress Cataloging-in-Publication Data

Names: Ardely, Anthony, author.
Title: What if corals disappeared? / Anthony Ardely.
Description: New York : Gareth Stevens Publishing, [2023] | Series: Life without animals | Includes index.
Identifiers: LCCN 2021044564 | ISBN 9781538276280 (set) | ISBN 9781538276297 (library binding) | ISBN 9781538276273 (paperback) | ISBN 9781538276303 (ebook)
Subjects: LCSH: Corals–Juvenile literature. | Corals–Behavior–Juvenile literature. | Corals–Conservation–Juvenile literature.
Classification: LCC QL377.C5 A73 2023 | DDC 593.6–dc23
LC record available at https://lccn.loc.gov/2021044564

Published in 2023 by
Gareth Stevens Publishing
111 East 14th Street, Suite 349
New York, NY 10003

Designer: Rachel Rising
Editor: Kate Mikoley

Photo credits: Cover, p. 1 Madelein Wolfaardt/Shutterstock.com; p. 3, 4, 6, 8, 10, 12, 14, 16, 18, 20, 22-24 Igor Kyrlytsya/Shutterstock.com; p. 5 Volodymyr Goinyk/Shutterstock.com; p. 7 John A. Anderson/Shutterstock.com; p. 9 SARAWUT KUNDEJ/Shutterstock.com; p. 11 Rich Carey/Shutterstock.com; p. 13 Lotus_studio/Shutterstock.com; p. 15 lunamarina/Shutterstock.com; p. 17 Magic Orb Studio/Shutterstock.com; p. 19 Tunatura/Shutterstock.com; p.21 V_E/Shutterstock.com.

Printed in the United States of America

CPSIA compliance information: Batch #CSGS23: For further information contact Gareth Stevens, New York, New York at 1-800-542-2595.

CONTENTS

Boldface words appear in the glossary.

Cool Corals

Some corals look like colorful underwater plants. However, corals aren't plants at all—they're animals! In fact, corals are some of the most important animals in the sea. Without them, our oceans, and planet, would look pretty different.

Polyp Power

Coral **organisms** are called polyps. Some are only about 0.04 inches (1 mm) across. Others grow as big as 10 inches (25 cm) across. Coral polyps have soft bodies. However, they also have a hard **skeleton** made of **limestone**.

Rock-Star Reefs

Some coral polyps live by themselves. When many live together, their skeletons can connect to each other and form a **colony**. Over many years, colonies join together forming a line of coral near the top of the water called a reef.

Life in the Reef

Coral reefs are important **habitats**. They cover only 1 percent of the ocean floor but are home to about 25 percent of the world's ocean creatures. Sea turtles, snails, and many fish make their homes in coral reefs.

Many plants also live in coral reefs. The organisms in a reef work together to create a healthy habitat. For example, plants give food and **oxygen** to the animals in the reef. Some animals keep the corals safe or clean the reef.

We Need Reefs

Reefs are helpful to more than just ocean organisms. They also help keep nearby land safe. Often near coasts, reefs take on some of the force from big waves and storms before they hit land. This can lessen the harm done to the land.

Coral may also be helpful in the field of health care. Scientists have found **chemicals** in corals that can help fight serious **infections**. Some coral skeletons can also be used in **surgeries** to take the place of human bone!

Corals in Trouble

Many reef-building corals are at risk of dying out. **Pollution** and rising ocean **temperatures** are just two of the ways they're being harmed. If corals disappeared, many animals would lose their homes and communities around coasts could be seriously hurt.

Save the Corals!

One way to help corals is to never put harmful chemicals down the sink. Even if you don't live near a reef, your water may end up by one. Keep learning about corals. Share what you know with friends and family.

GLOSSARY

chemical: matter that can be mixed with other matter to cause changes

colony: a group of animals living and working together

habitat: the natural place where an animal or plant lives

infection: the spread of germs inside the body, causing sickness

limestone: rock formed from skeletons and shells of sea creatures

organism: a living thing

oxygen: a colorless, odorless gas that many animals, including people, need to breathe

pollution: the process of making things such as land, water, or air dirty and unsafe to use

skeleton: a strong frame that supports an animal's body

surgery: a medical treatment used for injuries and illnesses that involves operations

temperature: how hot or cold something is

FOR MORE INFORMATION

BOOKS

Stefoff, Rebecca. *Corals: Secrets of their Reef-Making Colonies.* North Mankato, MN: Capstone Press, 2019.

Topacio, Francine. *Creatures of a Colorful Coral Reef.* New York, NY: PowerKids Press, 2020.

WEBSITES

Corals
www.dkfindout.com/uk/animals-and-nature/jellyfish-corals-and-anemones/corals/

Get an up-close look at corals and find more fun facts about these animals here.

Ocean Habitat
kids.nationalgeographic.com/nature/habitats/article/ocean

Read about several different ocean habitats, including coral reefs, on this website.

Publisher's note to educators and parents: Our editors have carefully reviewed these websites to ensure that they are suitable for students. Many websites change frequently, however, and we cannot guarantee that a site's future contents will continue to meet our high standards of quality and educational value. Be advised that students should be closely supervised whenever they access the internet.

INDEX